The Cooperative Learning Activities Handbook

by
Donna Boxx
Meri Makovec
Susie Zenner

Inside Illustrations by
Liza Sernett
based on the original work of Susie Zenner

Cover Illustration by
Lynda Lawson

Publishers
T.S. Denison & Company, Inc.
Minneapolis, Minnesota 55431

K. Campbell

I

T.S. DENISON & COMPANY, INC.

Standard Book Number: 513-02014-4
The Cooperative Learning Activities Handbook
Printed in the United States of America

Minneapolis, Minnesota 55431

CONTENTS

Acknowledgements . . .

Special thanks to our husbands, Chuck Zenner and David Makovec and to our families for their encouragement and support.

Who . . .

The Cooperative Learning Handbook offers suggested learning activities for the younger elementary school child - *Kindergarten through Second Grade.* However, cooperative learning can be successfully implemented with any age level through adulthood. Cooperative learning is designed to be used in heterogenous groups where children may learn from and support one another.

What . . .

Cooperative learning takes place when students work cooperatively in small groups on a multifaceted learning task. It is a strategy that can be used in combination with, or as an alternative to competitive and individualized learning formats.

When . . .

The cooperative learning format can be used whenever a learning goal lends itself to using higher level thinking skills and calls upon students to complete several tasks in order to attain that goal. Be sure to allow enough time for students to conclude the learning activity and also to discuss how effectively the groups have functioned.

Where . . .

Cooperative learning may be implemented in any classroom - indoors or out. Just make sure that the group stations are set up at a sufficient distance from one another to ensure minimal distractions.

Why . . .

Cooperative learning allows children to explore, to experiment and to create in a small group setting. Social and communication skills are learned and practiced when group interaction is necessary to accomplish an objective. Face-to-face interaction allows group members to speak, listen, assist, question and make joint decisions about the learning task. These types of group interactions mean students are more actively engaged in learning. Working together makes learning fun!

How . . .

You may begin by putting your children into groups of approximately four. Each group will be heterogeneous to maximize the diversity of experience, talent and intellect which each group has a whole. Each group should retain the same members for a number of activities. Be careful to place students so that each group has an equal distribution of boys and girls, high and low achievers, passive and active, verbal and less verbal.

The first goal may be to allow each group to choose a group name. A discussion of rules, or norms, which might assist groups in functioning effectively must also be elicited from the class. Keep the rules simple, positive and limited in number. These rules may change or become more sophisticated as group work evolves. Our original list included:

1. Everyone helps.
2. Everyone gets a turn.
3. Everyone listens.
4. Use "inside" voices.

The teacher's role in cooperative learning is that of a manager. The teacher sets up materials at the group stations in advance and is the source of ideas and solutions while group activities and procedures progress. The teacher must make clear the objectives and procedures for the cooperative learning task before the class breaks up into groups. At this point, the teacher oversees, rather that interferes in the group learning.

After the learning activitiy is concluded, the teacher may call the groups together and process information about how well the groups functioned. This is a good time to discuss any problems encountered in the group work and to reinforce the social and communication skills necessary for cooperative learning. A delegate from each group may also present the product of his/her group's work to the class. Feedback may be given at this time.

We think that students and teachers alike will find their venture into cooperative learning to be a rewarding and exciting experience. Students who are actively engaged in learning, creating and problem-solving are experiencing real academic growth. And the social and communication skills learned by the students during cooperative learning activities offer opportunities for social and emotional growth.

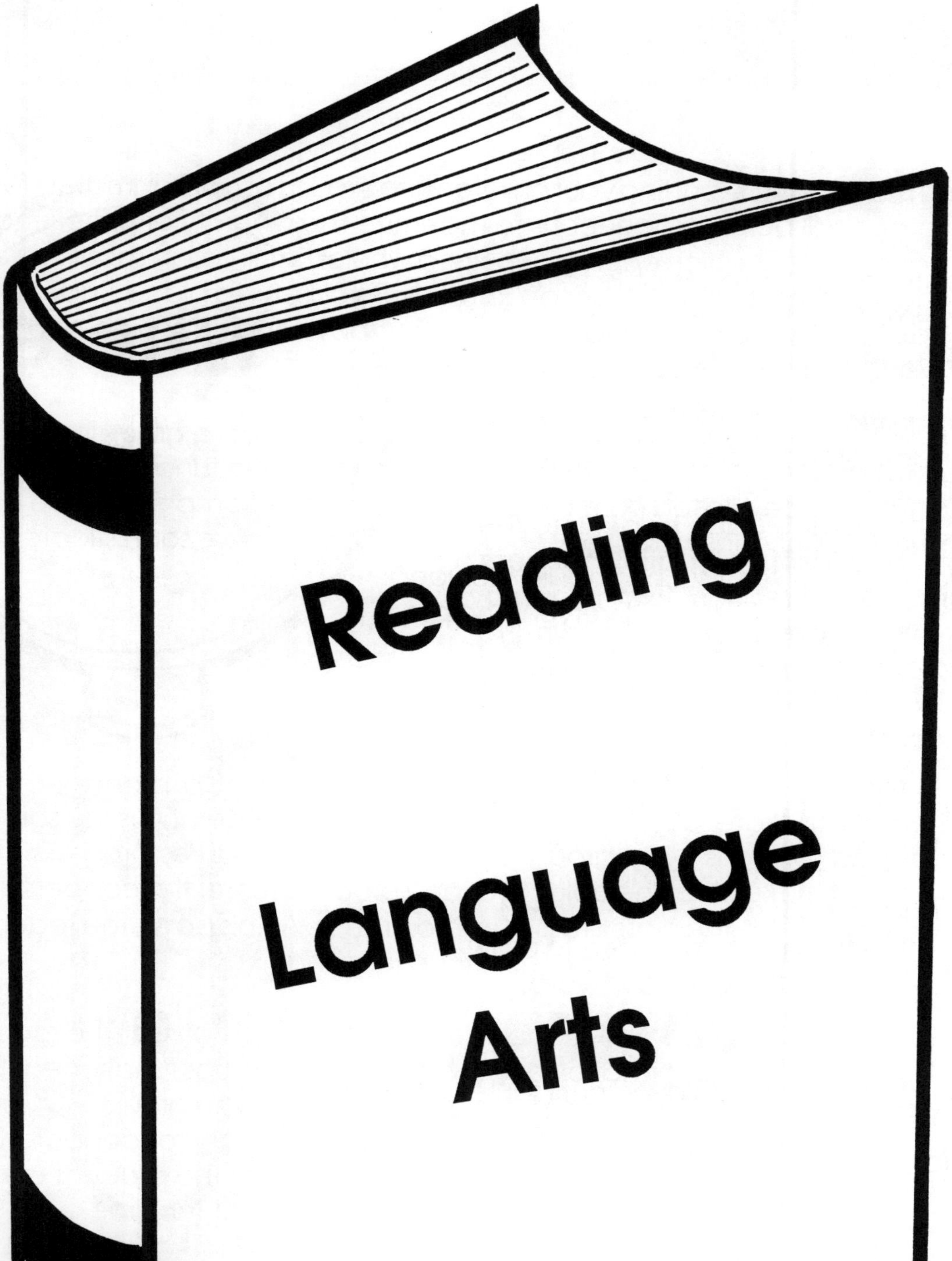
Reading
Language
Arts

Purpose
The students should be able to name and illustrate the given spatial concepts.

Preparation
At each group station provide a paper that has been prepared by the teacher. Divide each paper into four parts and print a "spatial concept" word in each section. Examples may be: over, under, in, out, up, down. Provide pencils, crayons and markers.

Procedures
1. Introduce the activity by reviewing the spatial terms to be utilized in the lessons.
2. At each station read the "spatial concepts" words to be illustrated.
3. Instruct the group members to draw a picture of the concept word.

Note
This lesson may be delayed until some of the group members are able to read.

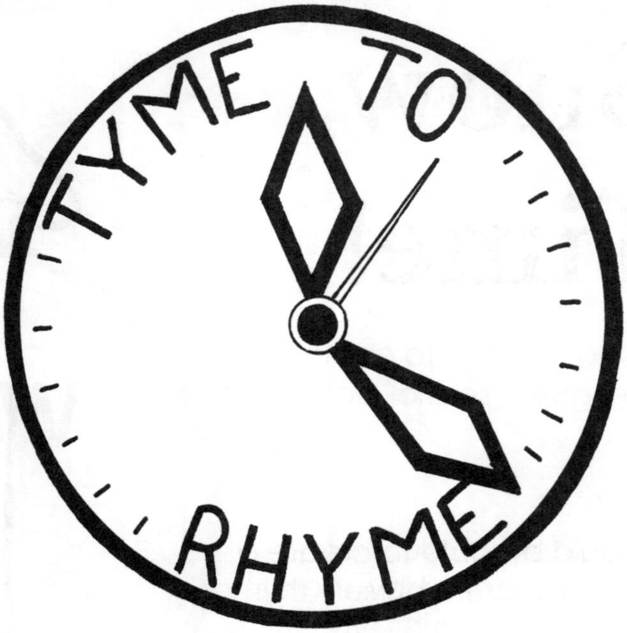

Purpose
Given a specific vowel-consonant ending, the students should be able to add beginning consonants to create rhyming words.

Preparation
Fold and prepare a large sheet of construction paper for each group. At the top of each paper print a vowel-consonant ending, such as _at, _ed or _in. Fold the paper in four or more equal parts. Print the vowel-consonant ending in each section. Provide pencil and crayons at each group station.

Procedure
1. Review the concept of rhyming with the class, pointing out the importance of the same ending sounds.
2. Introduce a sample paper prepared by the teacher for the class practice lesson.
3. Instruct students to name a beginning sound that will make a word by adding the letter to the vowel-consonant ending provided. Continue adding beginning letters until each section has a rhyming word.
4. Hand each group a prepared paper instructing the members to complete the activity as directed and continue by drawing a picture in each section above the rhyming word.

Note
A group member may be chosen to "present" the rhyming chart to the class.

Show Time!

Purpose
The students should be able to create a pantomime to dramatize a nursery rhyme.

Preparation
None

Procedure
1. Introduce the activity by reading a book of nursery rhymes.
2. Instruct each group to decide which nursery rhyme they will pantomime and to prepare it to present to the class.

Note
Let the class guess the nursery rhyme which the group presents.

Purpose
The students should be able to create a "little book" relating to a school activity, (field trip, guest speaker, enrichment activity, etc...)

Preparation
At each group station provide paper, crayons, markers and pencils.

Procedures
1. Introduce the activity by reviewing the experience from which the students will create a "little book."
2. Instruct each group to decide what will be included in their "little book."
3. Instruct each group to choose illustrators and writers for each page of the book.
4. Instruct each group to help all members "read" their little book.

Note
A group member may be chosen to present the "little book" to the class.

JUST THE OPPOSITE

Purpose
The students should be able to state and illustrate the opposite of a given word.

Preparation
At each group station provide a paper for each member which has been folded in half. Print a word which has an opposite on one half of the paper. (Examples; hot, up, tall, high, long, etc.)

Procedure
1. Introduce the activity by reviewing the concept of opposities and providing examples.
2. Instruct each group to help each member read his/her word, illustrate it and illustrate the opposite of the word.

Note
Each group may staple their papers together to form a booklet of "opposites" and share it with the class.

Purpose
The students should be able to follow a sequence of directions to make a peanut butter and jelly sandwich.

Preparation
At each group station provide bread, butter, peanutbutter, jelly, plastic knives and napkins.

Procedure
1. Tell the class you are making sandwiches for the snack of the day.
2. Instruct the class on how to make the peanutbutter and jelly sandwich. Emphasize the order of instruction.
3. Have each group member make a sandwich following the teacher's instructions.

Note
This activity works well when introducing "Ss" to the children. Sandwiches are fun "Ss" treats for the children to make and eat.

Tell Me About It!

Purpose
The students should be able to retell a story by creating a sequence of pictures.

Preparation
At each group station provide paper, markers and pencils.

Procedure
1. The class will listen to a story.
2. Instruct each group to retell the story and illustrate the important events in sequential order.
3. Paste the pictures on a large sheet of paper in proper order or staple the pages in booklet form.

Note
Children who are able to print sentences will enjoy adding sentences under each of the illustrations.

Purpose
The students should be able to create sentences to form a story about a given picture.

Preparation
At each station provide a set of pictures, writing paper and pencils.

Procedure
1. Introduce the activity by showing a picture to the class and eliciting sentences to create a story about the picture.
2. Instruct each group to choose a picture from the set.
3. Direct each student to create and add a sentence to his/her group story.
4. Select a "writer" for each group.
5. Instruct each group to help all members "read" the story.

Note
A group member may be chosen to read the story for the class.

Write On

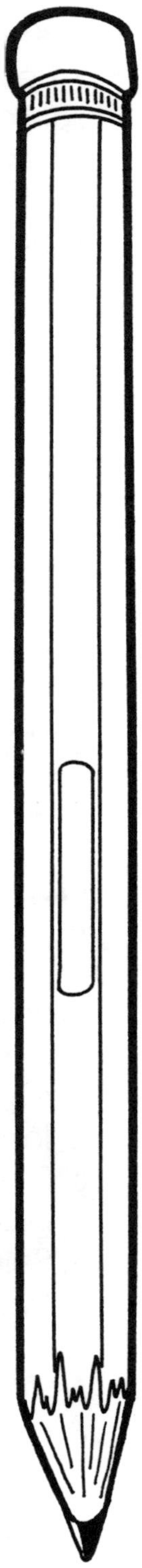

Purpose
The students should be able to write a story following a "What If" situation.

Preparation
At each group station provide paper and pencil.

Procedure
1. Introduce a situation to the whole class. Ex: "What if Santa couldn't get down the chimney?"
2. Instruct each group to discuss the situation and write a story telling about it.
3. A designated recorder may print the words as the group creates the story.

Note
A group member may be chosen to present the story to the class.

Purpose
The students should be able to find and cut out pictures from magazines that begin with a designated consonant sound.

Preparation
At each group station provide a large piece of paper or tagboard, scissors, paste and a pencil. Several old magazines should be placed at each station.

Procedure
1. Introduce the initial consonant sound to be reviewed by the class. Print the letter on the board and instruct group members to print it at the top of the tag board provided at each station.
2. Show the students how to search through the magazines cutting out all the pictures that begin with the designated consonant sound.
3. Instruct group members to paste all the pictures on the tagboard in collage fashion.

Note
Older children may wish to find pictures that have the same vowel sounds, consonant blends or ending sounds.

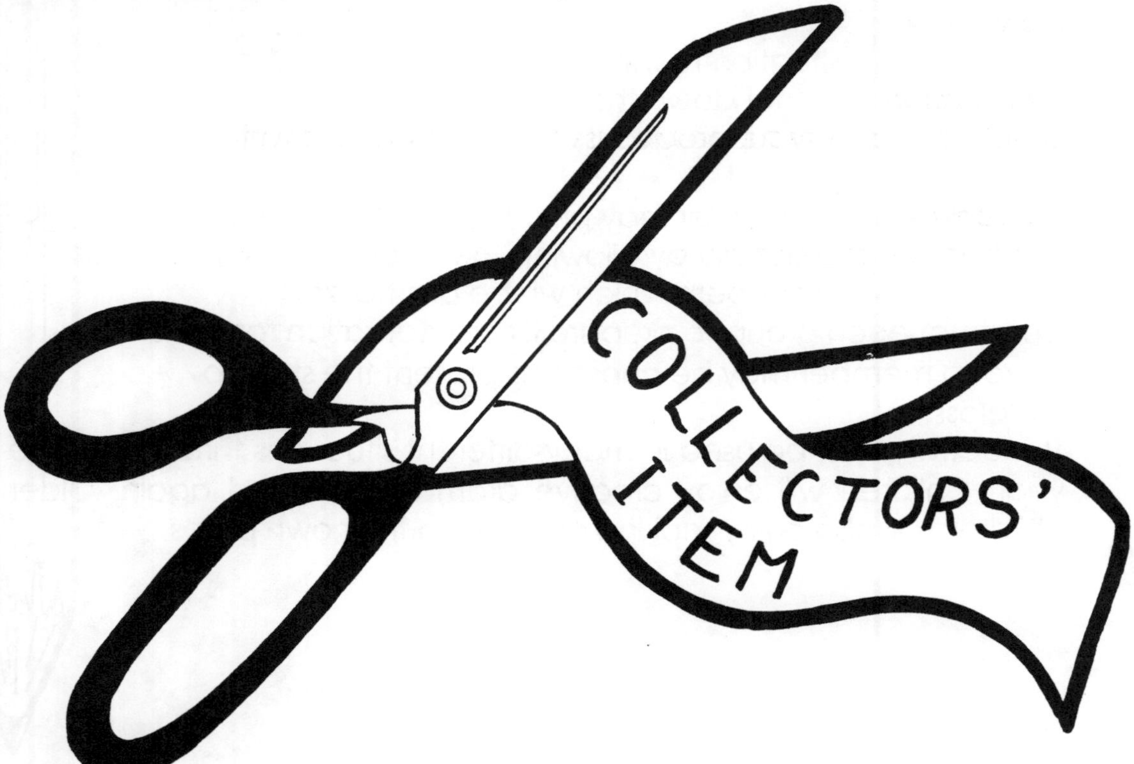

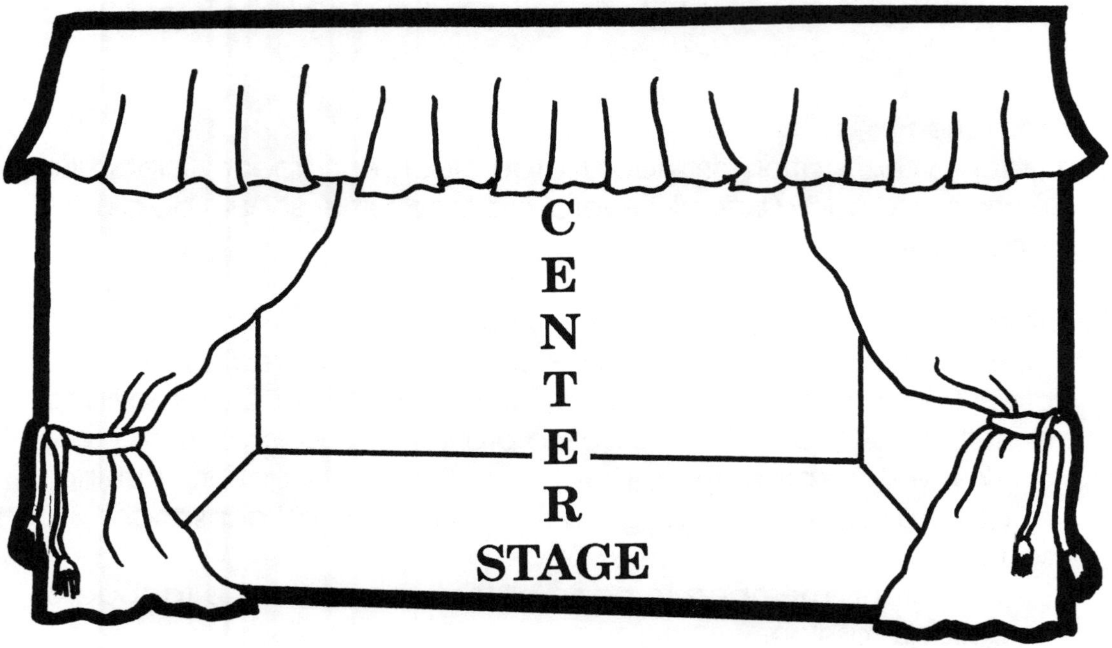

Purpose
The students should be able to dramatize a poem, fairy tale or other story.

Preparation
Simple props may be provided at each group station.

Procedure
1. Introduce the activity by allowing the students to hear and/or view the story or poem which will be dramatized.
2. Instruct each group to prepare its dramatization for the class.

Note
This activity can be used in many different situations throughout the year. Children will enjoy creative drama again and again. Older children will also enjoy making or creating their own props.

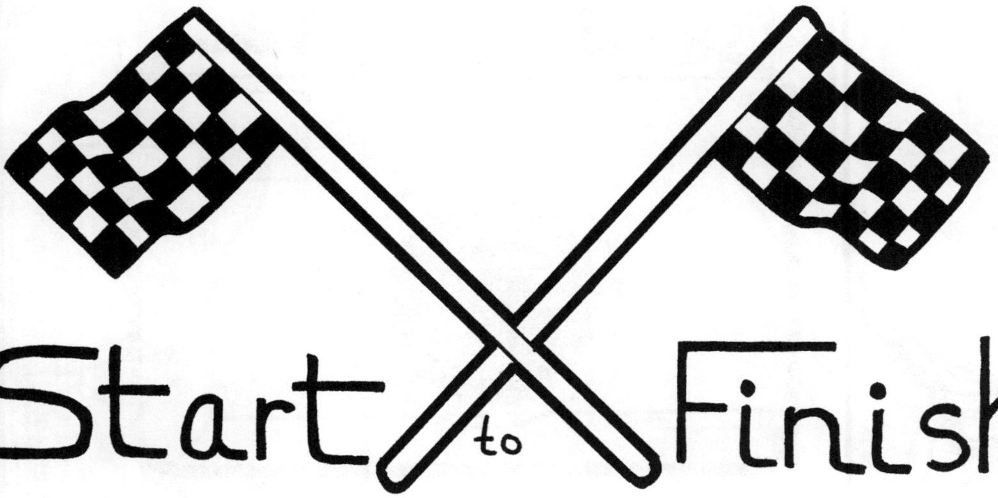

Start to Finish

Purpose
The students should be able to arrange picture cards in 1-2-3 order.

Preparation
At each group station provide sets of picture cards which can be arranged in 1-2-3 order to tell a story.

Procedure
1. Introduce the activity by sequencing a set of picture cards with the class.
2. Instruct each group to sort and sequence the cards provided at the station.

Note
Each group member may select a sequence and tell the story.

PICTURE THIS!

Purpose
The students should be able to create a mural by drawing pictures to depict a popular holiday story.

Preparation
At each group station provide a sheet of large rolled paper, markers, pencils, crayons and other creative materials.

Procedure
1. The teacher will read and discuss a popular holiday story with the class. Ex: The Night Before Christmas; Tommy Turkey; Peter Rabbit.
2. Instruct each group to create a mural by drawing pictures from the story. Begin on the left side with a picture showing what happened first in the story, pictures depicting what happened in the middle of the story, and a picture showing how the story ended.

Note
This is also an excellent activity to use when studying various stories from your classroom's reading program.

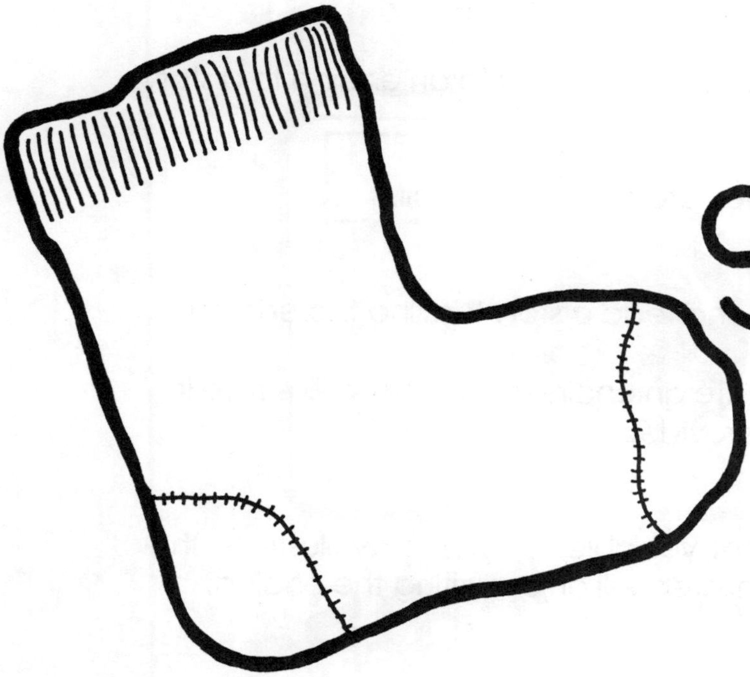

Sock It To Me.

Purpose
The students should be able to create puppets of community helpers.

Preparation
At each group station provide heavy paper, markers, crayons, glue, tongue depressors and scissors.

Procedure
1. As a part of the "community helpers" unit discuss the various jobs people do to help other people in the community.
2. Instruct group members to work in groups of two creating a "helper" of their choice. (Teacher may wish to assign occupations.)
3. Draw, color, cut and paste the "helper" on the tongue depressor for use in the puppet theatre.

Note
Any extra material available may be used to enhance the creative experience. Sock puppets work well too.

Purpose
The students should be able to create and write an ending to a story.

Preparation
Provide a large sheet of paper, crayons and pencil.

Procedure
1. The teacher should select and read a story leaving the ending open - to the entire class.
2. Instruct each group to create an ending to the story, illustrate it and write it on the paper provided.

Note
Younger children who are not yet able to read may illustrate their ending for the story. Older children will enjoy writing the ending.

FINISH UP

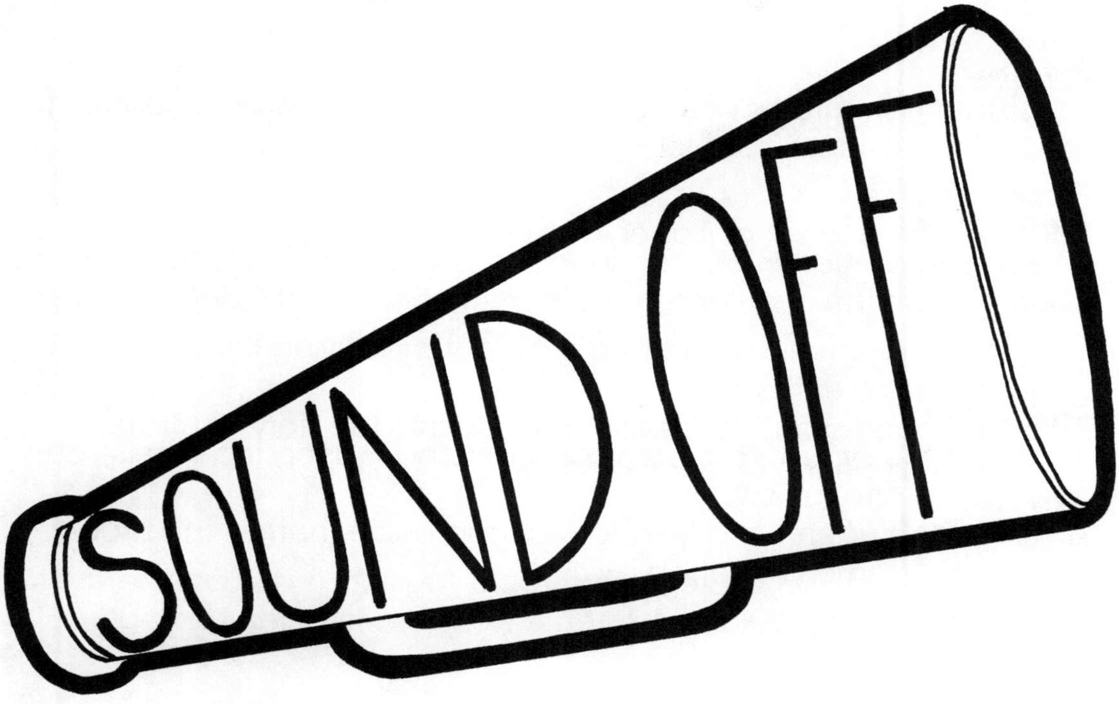

Purpose
Given a word with the initial consonant omitted, the students should be able to identify and print the missing letter.

Preparation
Prepare a chart with pictures of simple words. Print the words beneath the pictures, leaving a blank for the omitted initial consonant. Provide a chart at each group station.

Procedure
1. Introduce the activity by providing several examples similiar to the chart practice.
2. Instruct each group to complete the chart by printing the missing letter.

Note
Older children may enjoy creating the charts by themselves and then give the chart to another group to complete. Instead of using initial consonant sounds, vowels or other letters may be omitted. Older children may also wish to use this activity for enrichment work for their spelling words.

Purpose
The students should be able to match beginning consonant sounds of magazine pictures to the letter.

Preparation
At each group station provide magazines, glue and large paper for a chart. Create the chart using the letters or phonetic rules that you wish to review.

Procedure
1. Instruct the groups to create the letter charts using the letters you want them to review.
2. Have the students find the pictures and paste them in the proper letter column on the chart.

A PERFECT MATCH

Purpose
The students should be able to match upper and lower case letters of the alphabet.

Preparation
The teacher will prepare and provide a set of laminated ice cream cone match cards for each group.

Procedure
1. Review the proper sequence of the alphabet with the class.
2. Instruct group members to match the ice cream with the lower case letter to the cone with the upper case letter. Place the cones in sequential order from "Aa" to "Zz."

Imagine That

Purpose

The students should be able to create a picture and/or story depicting him/herself as a Pilgrim on the Mayflower.

Preparation

At each group station provide paper, pencil, crayons, paints and/or markers.

Procedure

1. Introduce the activity after providing sufficient information on Pilgrim life as part of a Thanksgiving unit.
2. Tell children, "You are a Pilgrim! What did you do on the Mayflower?"
3. Instruct each group to create a story or a picture to answer the question.

Note

This activity may be adapted to other situations where the students can imagine they are someone else or a character from a book.

ART

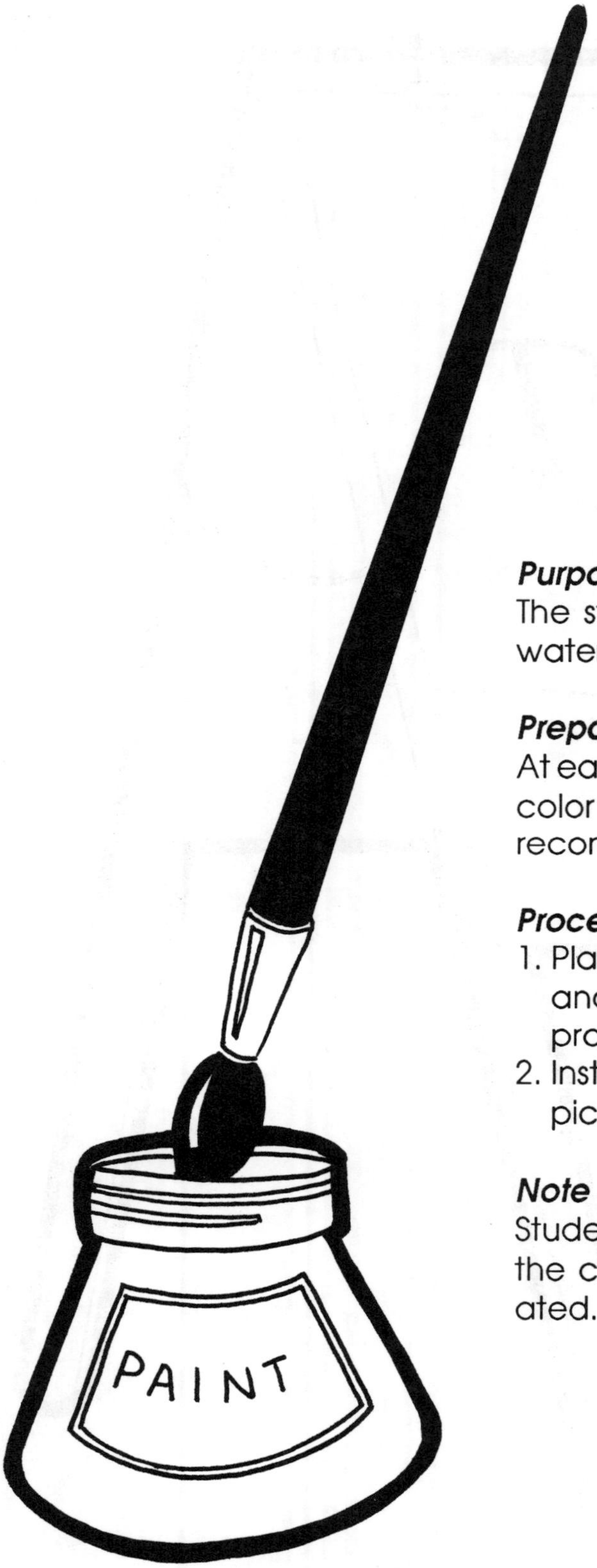

Busy Brushes

Purpose
The students should be able to paint with water colors.

Preparation
At each group station provide paper, water-color paint, brushes and water. Choose a record to play for a musical background.

Procedure
1. Play a portion of the record for students and discuss the feelings the music might produce.
2. Instruct each group to paint a creative picture while listening to the music.

Note
Students may enjoy showing their pictures to the class and telling about what they created.

Haunted Haven

Purpose

The students should be able to create a "spooky" Halloween picture.

Preparation

At each group station provide a large paper, pencils, crayons, markers and/or paints.

Procedure

1. Instruct each group to create a "spooky" Halloween picture after a discussion about Halloween.

Note

This activity may be adapted to any holiday!

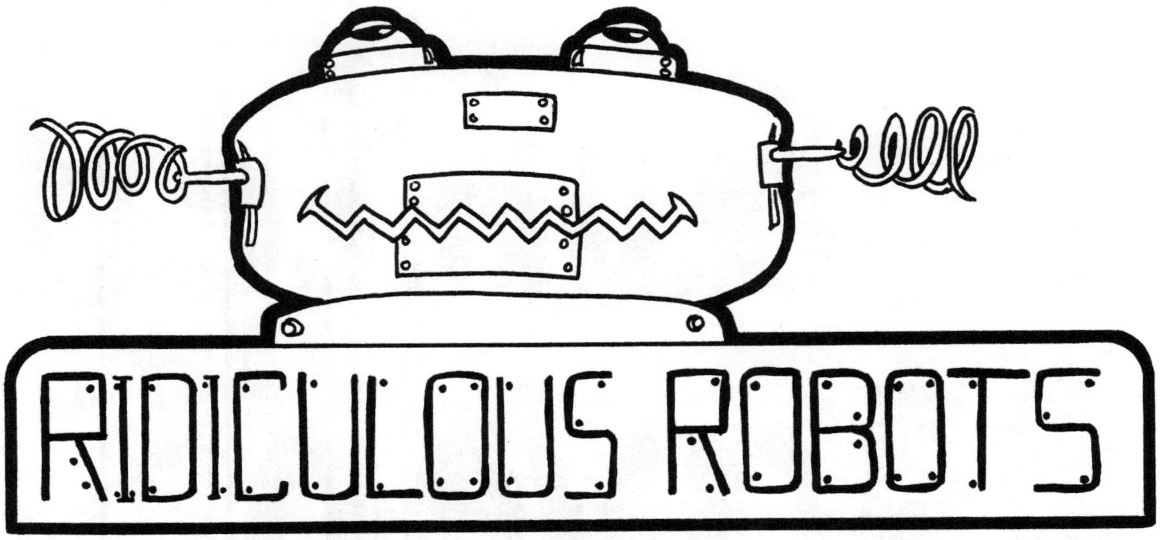

Purpose
The students should be able to create a "Ridiculous Robot" in conjunction with the study of the consonant "Rr."

Preparation
At each group station provide construction paper scraps of all colors and sizes, scissors, pencils and paste.

Procedure
1. Introduce the activity with a discussion of the terms "ridiculous" and "robot."
2. Instruct each group to create a "ridiculous robot" with the materials at the station and to be prepared to tell what the robot does.

Note
You may wish to have each group write a story about its "ridiculous robot." The children might also enjoy making a tape recording of things that the robot might say.

Purpose
The students should be able to draw a portrait of a student in their group.

Preparation
At each group station provide paper, pencils and crayons.

Procedure
1. Discuss the various facial features perhaps using volunteers from the class as examples.
2. Instruct each group to draw a portrait of someone in the group and color it. Encourage students to help each other by giving suggestions to one another.

Note
Students will enjoy playing "Guess Who" with the portraits they have drawn.

Purpose
The students should be able to create a banner or flag.

Preparation
At each group station provide paper, glue, scissors and construction paper scraps.

Procedure
1. Introduce the activity after a discussion of our flag as a symbol.
2. Instruct each group to create a banner or flag as the group's symbol.

Note
Banners may be made as a follow-up activity at various times of the year.

PLEASE REPEAT

Purpose

The students should be able to use a repeated pattern to create an Indian design for a headband or table runner.

Preparation

At each group station provide paper suitable for the project and pencils, crayons, markers and paints.

Procedure

1. Introduce the activity by showing examples of Indian designs and discussing the repeated patterns.
2. Instruct each group to create a design and to decorate a tablecloth or Indian headbands.

Note

The tablecloth or headbands might be used for a Thanksgiving Feast.

Purpose
The students should be able to express an idea and create a poster about it.

Preparation
At each group station provide large constuction paper or poster paper, markers, glue, pencils, crayons, scissors and other creative materials.

Procedure
1. Discuss with the class the theme for the poster project.
2. Instruct each group to create the poster showing an idea from the discussion.

Note
"How To" projects work well in lower grades. Examples: Care for pets; Be a good student for National Education Week; Care for teeth; Prevent fires.

Purpose
The students should be able to create a collage of natural items gathered on an autumn or spring outing.

Preparation
At each group station provide paper and paste.

Procedure
1. In conjunction with a fall or spring unit instruct the students to collect "sign of fall" or "signs of spring" while on a class outing.
2. Instruct group members to assist one another in creating a collage of the items collected.

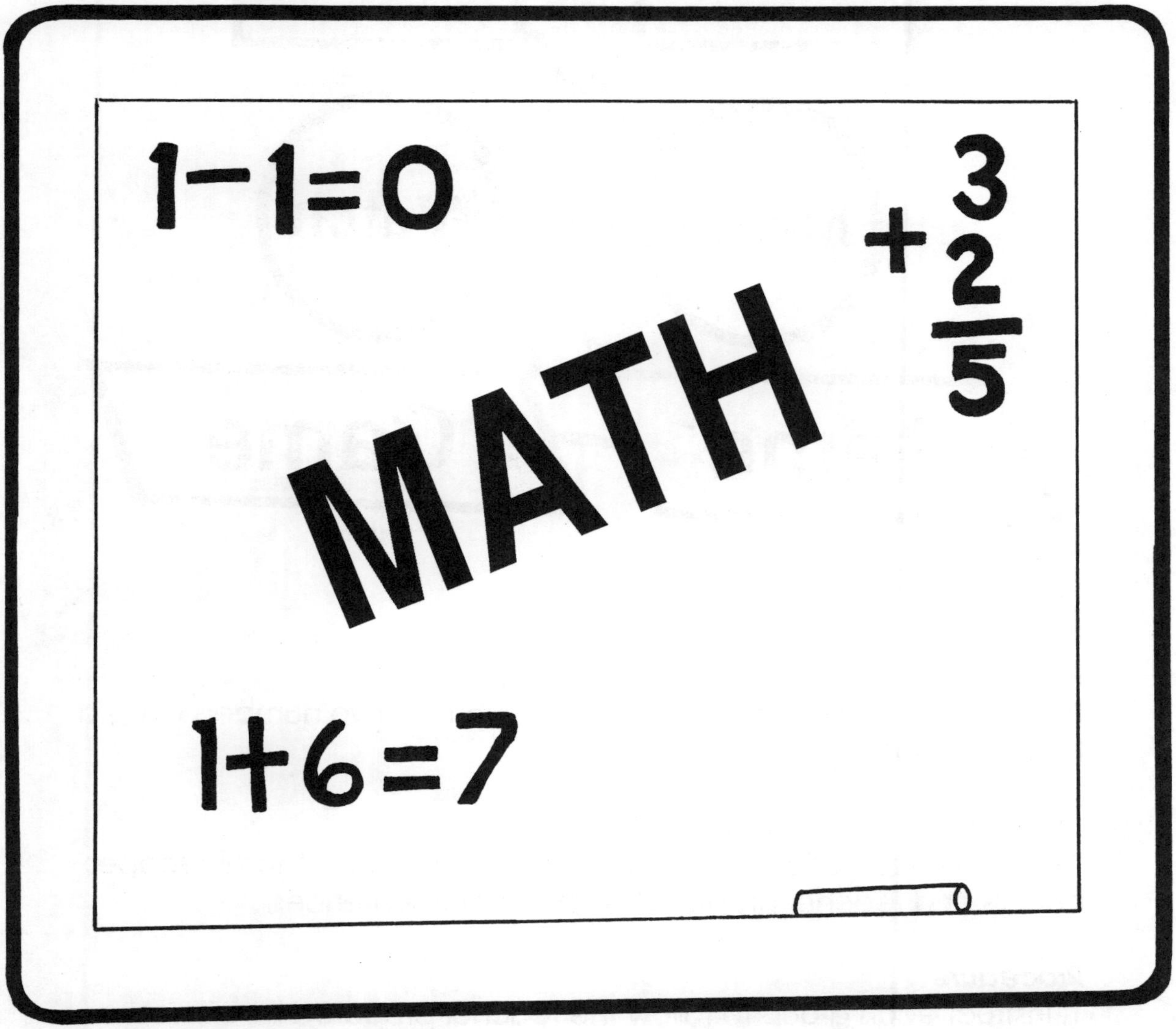
1−1=0
+ 3
2
5
MATH
1+6=7

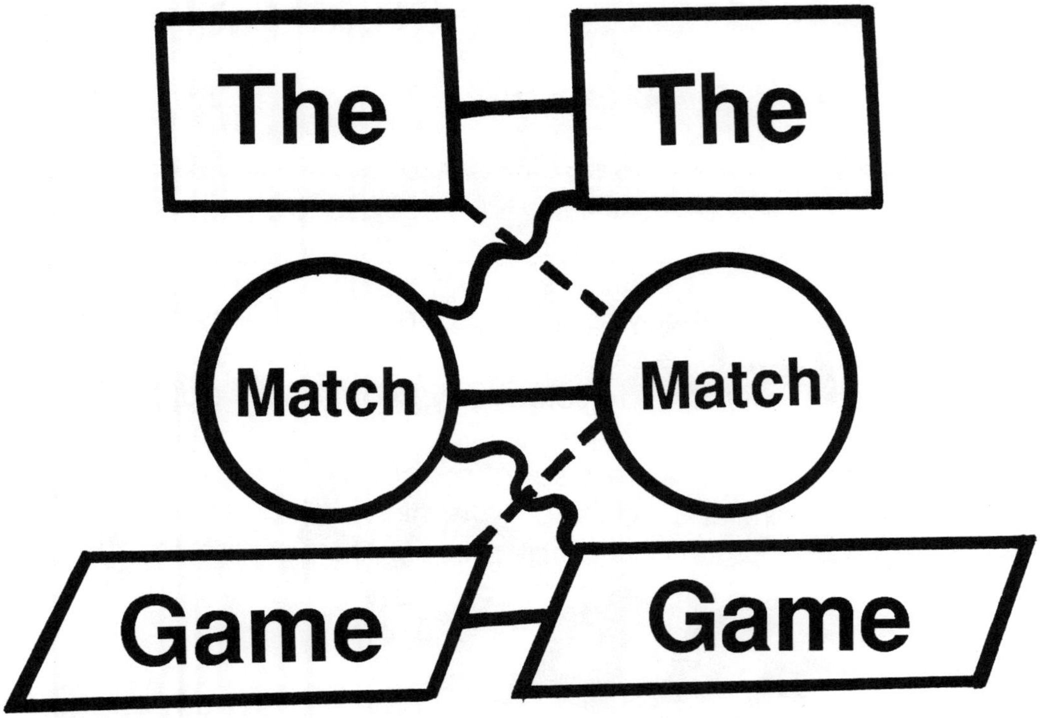

Purpose
The students should be able to add or subtract two numbers using a tachistoscope.

Preparation
At each group station provide several teacher made tachistoscopes, pencils and teacher prepared list of number sentences.

Procedure
1. Instruct each group to follow the teacher prepared list of number sentences. Move the tachistoscope windows to read the same as the problem on the list.
2. Have the group members arrive at an appropriate answer in using the tachistoscope and then print the answer in the answer box on the teacher prepared list.

Purpose
The students should be able to determine which of two numbers represents the greater amount.

Preparation
At each group station provide a pack of number cards and a bowl of macaroni. Provide paper and pencil for recording.

Procedure
1. Introduce the activity by reviewing concepts of "less than" and "greater than" with the class.
2. Instruct group members to select any two number cards from the deck of cards.
3. Tell the students to create a set of macaroni to represent each of the chosen number cards and determine which set has more.
4. Record the results on a chart by printing the numeral in the proper column.

Note
Older children may wish to use two and three digit numerals for this activity and eliminate the use of macaroni.

LESS	GREATER
2	6
4	8
1	3

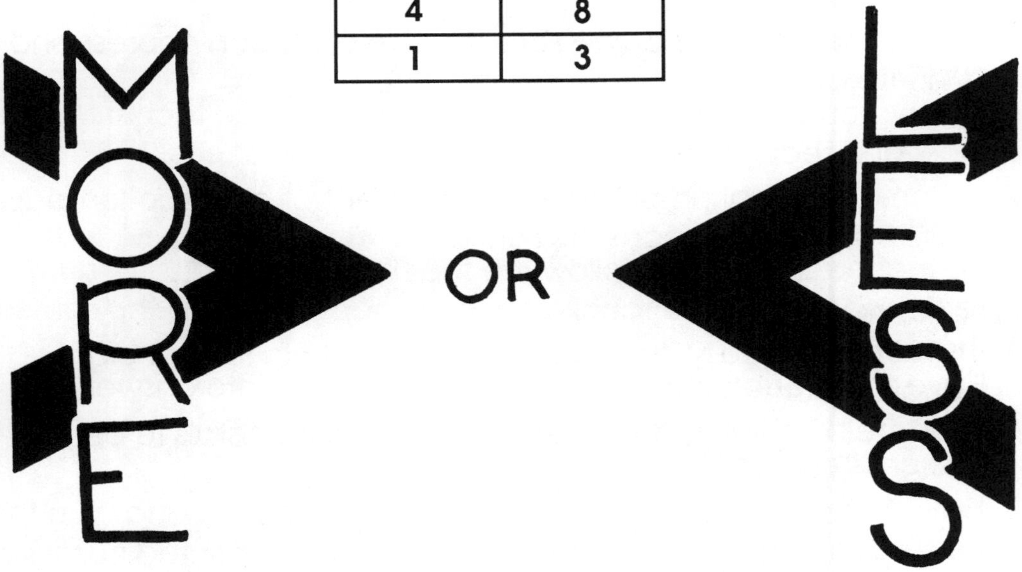

THIS COUNTS!

Purpose

The students should be able to create a set of objects cooresponding to a numeral.

Preparation

At each group station provide a large sheet of construction paper, pencil, macaroni or beads and glue.

Procedure

1. Instruct each group to fold the paper into two, four, or eight equal sections.
2. Indicate by printing on the chalkboard the numerals to be reviewed.
3. Have the groups print one numeral in each section and glue the corresponding amount of macaroni above the printed numeral.

NUM + BER = PICS

Purpose
The students should be able to draw pictures to illustrate number sentences.

Preparation
At each group station provide strips of paper, crayons and pencils. With marker, print a number sentence on each strip.

Procedure
1. Introduce the activity by reviewing examples of addition and/or subtraction sentences.
2. Instruct each group to illustrate the number sentences by drawing a set to show "how many" each numeral represents.

Note
You may print the entire number sentence or ask students to provide the sum or remainder.

3 + 2 = 5

Purpose
The students should be able to identify numerals 1 to 10 and draw sets to show "how many."

Preparation
At each group station provide a large sheet of paper, crayons and pencil. With marker, make ten large squares and print a numeral from 1 to 10 in each.

Procedure
1. Introduce the activity by reviewing the numerals 1 to 10.
2. Instruct each group to identify the numeral in each square and to draw a set to show "how many."

Note
This activity may be adapted to review any numerals taught.

Purpose
The students should be able to measure height and weight.

Preparation
At each group station provide a chart with a number line marked in inches and a scale. Also provide a card for each child to record his height in inches and weight in pounds.

Procedure
1. Introduce the activity with a discussion of how we measure height and weight. Include such terms as; inches, ruler, yardstick, scale and pounds.
2. Instruct each group to assist its members in measuring each child's height and weight and recording the facts on each individual's cards.

Note
Children might enjoy comparing their findings. Terms such as; tallest, shorter, heavier, lightest, taller, etc., could then be used.

Purpose
After reading a numeral, the students should be able to count the correct number of objects to make up the designated set.

Preparation
At each group station provide an empty egg carton with numerals 1 to 12 printed in the egg cups in the correct order. Provide counters such as pegs, and a set of cards with a numeral printed on each card.

Procedure
1. Review the concept of one-to-one coorespondence and counting as well as the numerals which will appear on the numeral cards.
2. Instruct each group member, in turn, to choose a numeral card. He will then put a peg in each cup, beginning with the number one until he reaches the numeral that appeared on the card. Other members will help to count, and check to see that the counters were correctly placed.

1st Things 1st

Purpose
The students should be able to identify the objects which are "first," "middle" and "last" in a row.

Preparation
At each group station provide a large paper, pencil, crayons, markers or paints.

Procedure
1. Introduce the activity by reviewing the concepts of "first," "middle" and "last."
2. In conjunction with a circus unit, for example, instruct each group to draw a circus parade.
3. Give specific oral directions to draw a lion first, an elephant in the middle and a clown last. Let the children create the remainder of the parade in any order.

Note
These will make great bulletin boards and the "parade" concept may be used in conjunction with a variety of units.

Purpose
The students should be able to add the two sets of numbers represented by the dots on a domino.

Preparation
At each group station provide a set of dominos, paper and pencil.

Procedure
1. Instruct group members to use one domino to create an addition number sentence.
2. Students will use the paper to print the number sentence and fill in the correct answer by counting all the dots on the domino.

Quarter Time

Purpose
The students should be able to divide objects into four equal parts and identify the parts as fourths or quarters.

Preparation
At each group station provide a scissors and serrated table knife for cutting. Provide objects for quartering, such as string, paper, an apple, a banana and a candy bar.

Procedure
1. Introduce the activity by defining the concept of quartering as dividing an object into four equal parts.
2. Instruct each group to quarter the objects provided at the stations.

Note
This activity may be adapted for teaching concepts of other fractional parts.

Purpose
The students should be able to add two numbers and transfer the numeral to a consonant code so a written message will appear.

Preparation
The teacher will prepare work sheets using addition number sentences with specific answers that may transfer to a consonant code. Example: The message to be decoded may be "GOOD JOB." The teacher must prepare the work sheet and provide the transfer information for each group.

Procedure
1. Instruct group members to complete the addition first.
2. Transfer the number answer to the consonanat code and write the letter in the appropriate message slot.

1+0+1	**1+1=2**	**2+0=2**	**1+2+3**	**2+2=4**	**1+1=2**	**2+3=5**
G	**O**	**O**	**D**	**J**	**O**	**B**

1=G 2=O 3=D 4=J 5=B

SCIENCE

Purpose

The students should be able to describe the tastes of a variety of foods.

Preparation

At each group station provide a tray with a variety of foods with different tastes; sweet, sour, salty, and bitter. Have an aide, parent or upper grade student at each station to assist with this activity. A blindfold will add to the fun!

Procedure

1. Introduce the activity by reviewing the sense of taste and several words which might be used to describe different tastes; sweet, sour, bitter salty.
2. Instruct each group member to take turns tasting, describing and guessing the food sample offered by the helper at her/her station.

Note

For easy serving, have the food samples bite-sized with a toothpick in each. Older children may enjoy charting their responses.

WEATHER WATCHERS

Purpose

The students should be able to observe and chart daily weather changes for one week. (Or longer than a week if you wish.)

Preparation

The teacher will need to prepare a chart for each group station. The days of the week can be listed at the top and ample space left for the students to draw or cut and paste the appropriate symbol of the weather for that day.

Procedure

1. Discuss with the class the types of weather you wish to consider and establish a symbol for each condition.
2. Instruct each group to appropriately display the symbol on the chart under the correct day.
3. Display the completed charts on a bulletin board for extended learning experiences.

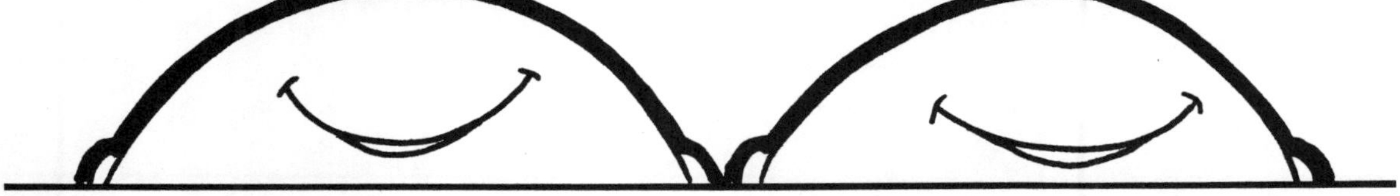

Purpose
The students should be able to create designs by blowing through a straw to spread paint.

Preparation
At each group station provide paper, liquid paint, eye dropper, pencil and straws.

Procedure
1. Instruct students to use the eye dropper to drop small amounts of paint on the paper.
2. Have the students immediately blow the paint in a variety of directions until a design is created to fill the page.

Note
One large paper may be used or each group member may do an individual creation.

Purpose

The students should be able to classify a variety of items by texture.

Preparation

At each group station provide paper, pencil, crayons, and a variety of items for texture classification.

Procedure

1. Discuss with the class the texture concept "How does it feel?" Determine several categories; rough, smooth, soft, etc.
2. Instruct the groups to feel and observe the items at the station, separating them into the categories discussed above.
3. Suggest each group use the paper as a recording tool to draw the items in sets according to "how they feel."
4. Select one member from each group to explain the groups sorting process.

TEXTURE TACTICS

Purpose

The students should be able to create a picture using the designated paper shapes.

Preparation

At each group station, place cut outs of the desired shape and size, large construction paper, paste and markers.

Procedure

1. Review with the large group the name of the various shapes chosen for this lesson.
2. Instruct each group to create pictures of objects or scenes by pasting the shapes on the large construction paper.
3. Marker may be used to add detail if you desire.

Note

This lesson can be repeated as you introduce each of the five shapes. Older children will enjoy doing this activity using more of the complicated shapes.

Purpose

The students should be able to observe the three states of water: liquid, solid and gas.

Preparation

At each group station provide a clear glass dish and paper for recording the data.

Procedure

1. Discuss with the class the three states of matter. Use water as an example.
2. Instruct each group to gather a dish of snow from outside. (Ice cubes may be used if snow is not available.)
3. Encourage each group to create a chart to record the amount of time it takes for the solid to return to liquid and then observe daily changes until the liquid has evaporated.

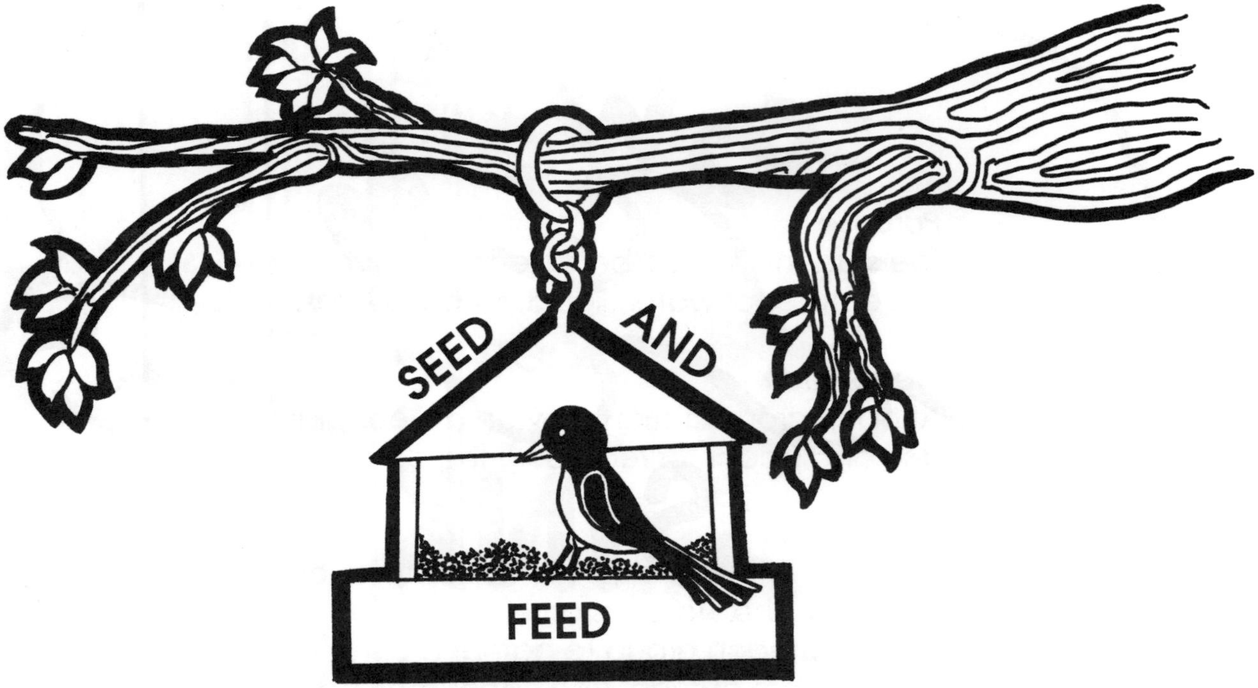

Purpose
The students should be able to construct a bird feeder.

Preparation
At each group station provide a large pine cone, peanut butter, bird seed, table knife and string.

Procedure
1. Discuss with the class the value of providing birds with food.
2. Instruct each group to create a feeder by spreading the peanut butter on the cone with the knife, rolling it in bird seed and tying a string on the top of the cone.
3. Provide an opportunity for each group to walk outside and hang its feeder in an appropriate place.

Note
This is a good project for early winter. The groups will enjoy monitoring the feeders to see which feeder is being used the most often by the birds.

Purpose
The students should be able to determine through experimentation the objects that will float and the objects that will sink.

Preparation
At each group station provide a container of water and a variety of small objects for experimentation purposes. Provide large paper, pencils, markers and crayons.

Procedure
1. Introduce each group activity by reviewing the concepts of "float" and "sink."
2. Show the groups how to prepare two observation charts using the large paper. Print the word "FLOAT" at the top of one and "SINK" at the top of the other.
3. Instruct each group to experiment with the various objects found at its station. If the object floats when placed in the container of water draw a picture of it on the observation chart marked "float." If the object sinks record it on the observation chart marked "sink."

Note
A group member may be chosen to present the chart to the class.

Purpose
The students should be able to separate animal pictures into three categories: zoo animals; farm animals, and; house pets.

Preparation
At each group station provide magazines, scissors, paste and chart paper divided into three sections labled, zoo animals, farm animals and house pets.

Procedure
1. Instruct each group to cut out animal pictures from the magazines provided.
2. Have the members paste each animal picture in the correct section on the chart paper.

Note
Older children may wish to write lists of these different categories of animals. A group member may be chosen to present the chart to the class.

TAKE ME HOME

Purpose
The students should be able to plant seeds and observe plant growth.

Preparation
At each group station provide bean seeds, soil, water and an allumminum pie pan or other container suitable for planting.

Procedure
1. Introduce the activity by reviewing the things which a plant needs to have in order to grow.
2. Demonstrate the procedure to be followed to plant the seeds and let the groups use their "green thumbs."

Note
You may wish to have the children observe and record the plant growth or by making a booklet. The children will also think it is great fun if you give each of the groups different seeds. Which group's seeds sprouted first? Can the groups guess what type of plants the other groups have planted?

Purpose
The students should be able to follow a simple recipe to prepare instant pudding.

Preparation
At each group station provide cooking utensils and ingredients needed to prepare the recipe.

Procedure
1. Read, discuss and show by example, if necessary, how to prepare the recipe.
2. Instruct each group to prepare the recipe. Supervise the preparing and enjoy the results!

Note
Any simple recipe may be used. The children will also think the pudding cooking experience is fun if each group is given a different flavor.

Purpose
The students should be able to determine through experimentation which objects will attract a magnet.

Preparation
At each group station provide a magnet and a variety of objects for experimentation. Provide large paper, pencils, markers and crayons.

Procedure
1. Introduce the activity by reviewing the "magnetic" concept.
2. Instruct each group to experiment with the magnet and various objects to determine which ones will attract.
3. Show the students how to tabulate the findings on a recording sheet.

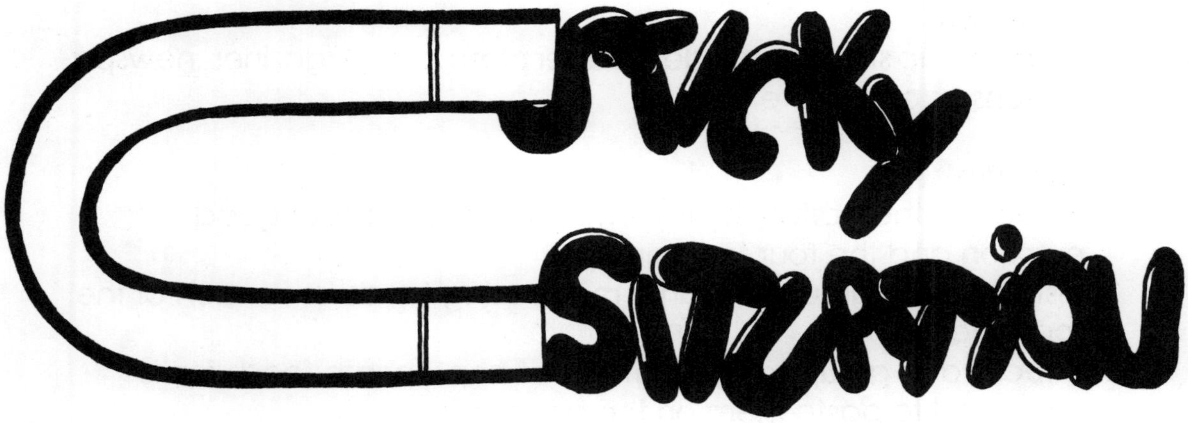

Purpose
The students should be able to plan a menu including the four food groups.

Preparation
At each group station provide a paper plate, old magazines, newspapers, scissors and paste.

Procedure
1. Introduce the activity by reviewing the concepts of good nutrition and the four food groups.
2. Instruct each group to plan a menu using foods from each of the food groups.
3. Direct each group to find pictures of appropriate foods for its meal and to paste them on the paper plate.

Note
A group member may be chosen to tell the class about the meal his/her group has planned. Foods may be drawn rather than cut and pasted if you desire.

Purpose
The students should be able to describe and draw various characteristics of animals preparing for winter.

Preparation
At each group station provide scissors, paper, paste, markers, pencils and crayons.

Procedure
1. Introduce this activity by creating a chart with ideas about how animals prepare for winter. (Humans may also be included.)
2. Instruct each group to select and create several ideas from the chart.
3. Display the group creations by arranging them on a bulletin board entitled, "Getting Ready For Winter."

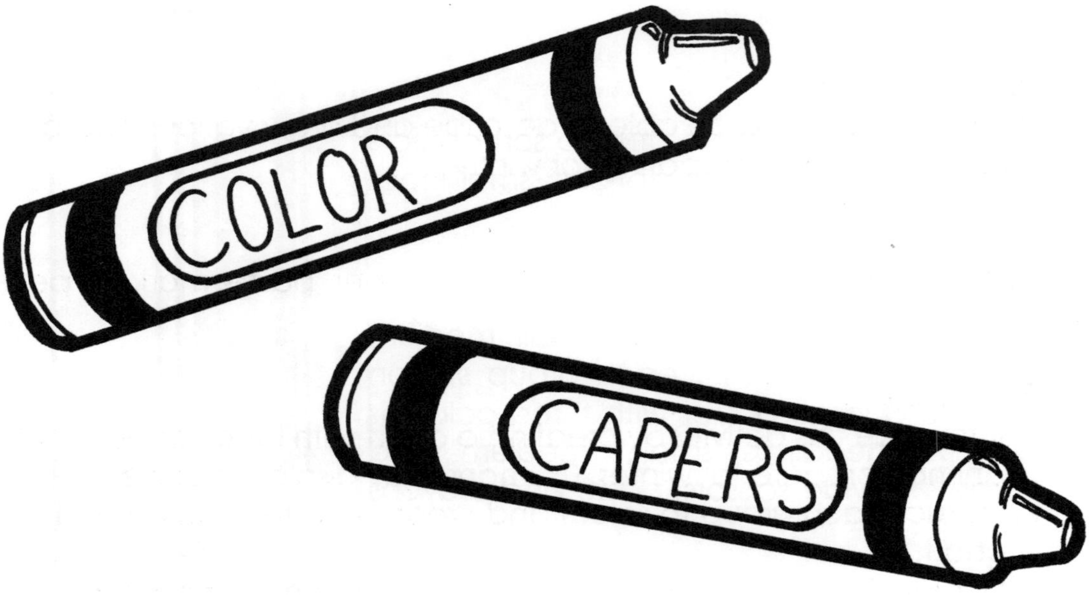

Purpose

The students should be able to classify pictures according to color.

Preparation

At each group station provide a large paper with columns for the colors students are to be classifying. Provide old magazines, catalogs, scissors and paste.

Procedure

1. Instruct each group to cut out pictures and paste them under the correct color heading.

Note

This is an excellent activity for a variety of classifying experiences. The children could look through catalogs and magazines for classifying objects for different rooms in a home; clothing; toys; silverware; foods; things made from wood; things made from glass; etc.

RED	BLUE	YELLOW

HATS	SOCKS	SHOES

Purpose

The students should be able to sort objects according to size descriptions; small, medium, and large.

Preparation

Prepare a ditto master with small, medium and large shapes. At each group station provide a large paper with three columns with the headings "small," "medium" and "large." Provide scissors, paste and crayons as well.

Procedure

1. Review the concept words small, medium and large.
2. Instruct each group to cut out the shapes and assist one another in pasting them under the correct headings. You may also wish for the children to color the shapes.

Note

You may want to use objects or other pictures to classify according to size.

The Four Seasons

Purpose
The students should be able to identify the four seasons and the changes brought about by each of the seasons.

Preparation
At each group station provide paper for booklets, pencils, crayons and markers.

Procedure
1. Review the concept of seasonal changes, perhaps using a picture book to aid in discussion.
2. Instruct each group to create a booklet illustrating an apple tree in spring, summer, fall and in winter. Use a page for each season and have the students print the name of the season on the page when the illustration is completed.

Note
Booklets or murals may be used to illustrate other seasonal changes such as; clothing, activities, or sports.

Social
Studies

Pilgrim Perils

Purpose
The students should be able to draw pictures showing conveniences we have today that the Pilgrims did not have.

Preparation
At each group station provide a large piece of paper, crayons, markers and pencils.

Procedure
1. As a part of the Thanksgiving Unit discuss the many conveniences that make our life easier than it was in the days of the Pilgrims.
2. Instruct each group to create a chart showing modern conveniences that make our life easier.

Note
A comparative chart may also be used. Students show how the Pilgrims made bread and how most people get bread today. How the wash was done then and now, etc.

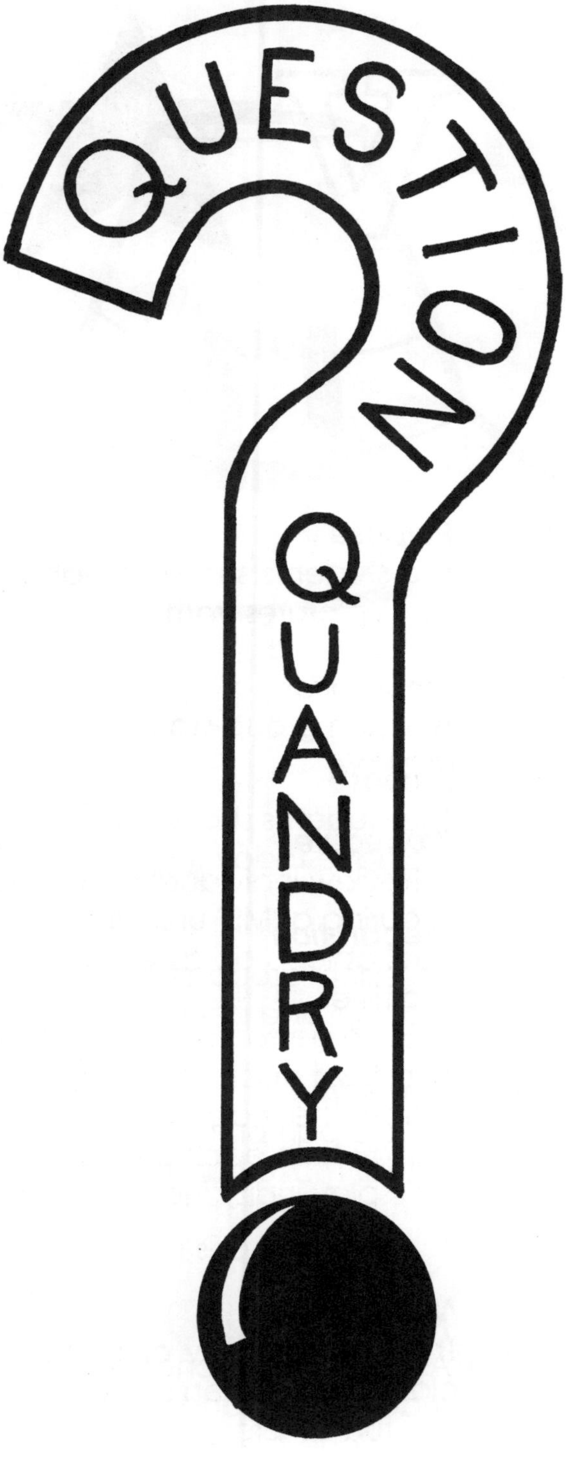

Purpose
The students should be able to prepare appropriate questions suitable for use on a field trip.

Preparation
At each group station provide a blank tape and tape recorder.

Procedure
1. Discuss with the class the difference between questions and answers.
2. Instruct each group to discuss and develop three questions appropriate for a specific field trip.
3. After checking with the teacher or assistant each group will select one person to record the questions on the tape.
4. The tapes will be collected and shared with all the class members.

Note
The fire station is a good example of a field trip for which group members may prepare their questions.

V★I★P

Purpose
The students should be able to create a collage about themselves using pictures from various magazines.

Preparation
At each group station provide magazines, paper, glue, scissors and markers.

Procedure
1. Review individual characteristics with the class as discussed during a "ME" unit. Stress that all of us are individuals, special are all able to do different things and all have different likes and dislikes.
2. Each person prints "ME" at the top of a paper.
3. Instruct each group member to create a collage using pictures from the magazines to describe themselves. Group members may help each others find appropriate pictures.
4. Display individual collages on a bulletin board during the "ME" unit week.

Note
The children may enjoy taking turns presenting the collages to the class. The children could either present their own "ME" collage or have the fun of presenting a collage made by one of the other group members.

T.V. TUBE

Purpose
The students should be able to create a television play and assemble a cardboard television to present the program.

Preparation
At each group station provide a roll of white shelf paper, two empty paper towel rolls, glue, markers, pencils and a box the width of the paper towel rolls.

Procedure
1. Discuss with the group several ideas for a play. (Mother Goose stories and rhymes work well. Groups may also create their own stories.)
2. Instruct each group to draw pictures on the shelf paper from top to bottom illustrating the story of the group's choice.
3. Roll the shelf paper on the empty towel rolls and attach to the inside of the box.
4. Cut out a window in the box and roll the pictures through as one member retells the story.

Purpose
The students should be able to act out various roles depicting community helpers.

Preparation
No special materials are necessary for this activity.

Procedure
1. As part of the "community helpers" unit discuss the various jobs people do to help other people in the community.
2. Instruct group members to take turns acting out a job performed by one of the community helpers.
3. Other group members will guess the occupation of the performer.

Note
The children may wish to find or make their own props for their performances.

TELL IT LIKE IT IS

Purpose
The students should be able to categorize pictures as "farm" or "city" settings.

Preparation
At each group station provide paper divided into two columns; city and farm. Also provide magazines, scissors and paste.

Procedure
1. Initiate a discussionaboutf the differences between "farm" and "city" settings.
2. Instruct each group to find magazine pictures depicting farm and city settings and place them in the proper category column on the paper. (The children may also draw the pictures.)

Note
Students may wish to dramatize or write about a farm or city based upon a picture they find.

Purpose
The students should be able to create a map of the classroom, school or playground.

Preparation
At each group station provide a large piece of paper, crayons and pencil.

Procedure
1. After a lesson on maps, instruct each group to create a map of the classroom, school or playground.

FACES AND FEELINGS

Purpose
The students should be able to create faces depicting various emotions.

Preparation
At each group station provide four paper plates, pencils crayons and markers.

Procedure
1. After a discussion of feelings, instruct each group to create faces that are happy, sad, frightened and angry.
2. Each group member may create a face from one of the paper plates. Each group member may decide which emotion they would like to illustrate.

Note
You may wish to provide sticks to be attached to the paper plates to make puppets. Students may wish to create a play using their puppets.

FUN IN THE SUN

Purpose
The students should be able to categorize summer activities and winter activities.

Preparation
At each group station provide paper, scissors, old magazines and paste. Divide the paper into sections: "Summer Fun" and "Winter Fun."

Procedure
1. In conjunction with a unit on the seasons, initiate a discussion of how people's activities differ in summer and in winter.
2. Instruct each group to find pictures for each of the two categories and paste them in the appropriate section.

Note
You may wish to have students draw pictures or write about the seasons instead of using magazines.